YOUR KNOWLEDGE HAS VALUE

Bradley Tice

Patterns within Pattern-less Sequences

GRIN Verlag

Bibliografische Information der Deutschen Nationalbibliothek:

Die Deutsche Bibliothek verzeichnet diese Publikation in der Deutschen National-
bibliografie; detaillierte bibliografische Daten sind im Internet über http://dnb.d-
nb.de/ abrufbar.

Imprint:

Copyright © 2012 GRIN Verlag GmbH
Druck und Bindung: Books on Demand GmbH, Norderstedt Germany
ISBN: 978-3-656-64525-2

This book at GRIN:

http://www.grin.com/en/e-book/195912/patterns-within-pattern-less-sequences

GRIN - Your knowledge has value

Der GRIN Verlag publiziert seit 1998 wissenschaftliche Arbeiten von Studenten, Hochschullehrern und anderen Akademikern als eBook und gedrucktes Buch. Die Verlagswebsite www.grin.com ist die ideale Plattform zur Veröffentlichung von Hausarbeiten, Abschlussarbeiten, wissenschaftlichen Aufsätzen, Dissertationen und Fachbüchern.

Visit us on the internet:

http://www.grin.com/

http://www.facebook.com/grincom

http://www.twitter.com/grin_com

Patterns within Pattern-less Sequences

Bradley S. Tice

Advanced Human Design, P.O.Box 3868 Turlock, California 95381 USA

While Kolmogorov complexity, also known as Algorithmic Information Theory, defines a measure of randomness as being pattern-less in a sequence of a binary string, such rubrics come into question when sub-groupings are used as a measure of such patterns in a similar sequence of a binary string. This paper examines such sub-group patterns and finds questions raised about existing measures for a random binary string.
PACS Numbers: 89.70tc, 89.20Ff, 89.70tc, 84.40Ua

Qualities of randomness and non-randomness have their origins with the work of von Mises in the area of probability and statistics [1]. While most experts feel all random probabilities are by nature actually pseudo-random in nature, a sub-field of statistical communication theory, also known as information theory, has developed a standard measure of randomness known as Kolmogorov randomness, also known as Martin-Lof randomness, that was developed in the 1960's [2,3& 4]. This sub-field of information theory is known as Algorithmic Information Theory [5]. What makes this measure of randomness, and non-randomness, so distinct is the notion of patterns, and pattern less, sequences of 1's and 0's in a string of binary symbols [6]. In other words, perceptual patterns as seen in a sequence of objects that can be defined as having similar sub-groupings within the body of the sequence that have a frequency, depending on the length of the string, of either regularity, non-randomness, or infrequency, randomness, within the sequence itself [7].

In examining the classical notion of a random and non-random set of 1's and 0's in two examples of a sequence of binary strings, the pattern verses pattern-less qualities can be examined.
Example #1 is as follows: [111000111000111] and Example #2 is as follows: [110111001000011]. It is clear than Example #1 is more patterned than Example #2 in

that Example #1 has a balanced sub-groups of three characters, either all 1's or all 0's, that have a perceptual regularity. Example #2 is a classical model of a sequence of a random binary string in that the sub-groups, if grouped into like, or similar, characters, either all 1's or all 0's like in Example #1, the frequency of the types of characters, either 1's or 0's, is different, seven variations of groups as opposed to the five variations in Example #1, as are the subgroups: [(11), (0), (111), (00), (1), (0000), & (11)] from Example #2. While this would support the pattern verses pattern-less model proposed by Kolmogorov complexity, there is a striking result from these two examples, #1 and #2, in that the second, or random, example, Example #2, has a pattern within the sub-groups, that for all perceptual accounts, has distinct qualities that can be used to measure the nature of randomness on a sub-grouped level on examination of a binary string.

The author has done early work on coding each of the sub-groups and reducing them to a compressed state, and then decompressing them with no loss to either the amount of frequency or number of characters to a sequence of a binary string that would be considered random by Kolmogorov complexity [8]. Now, while this simple program of compression and decompression by the author is for a future paper, the real interest of this paper is on the sub-groups as they stand without the notion of compression.

The very idea of the notion of a patterned or pattern-less quality as found in the measure of such aspects to the sub-groupings of 1's and 0's in a sequence of a binary string has the quality of being a bit vague, in that both Example #1 and Example #2 are patterned, in that they have a frequency and similar character sub-groupings that have a known measure and quality that can be quantified in both examples. This is more than a question of semantics as the very nature of the measure of Kolmogorov complexity is the very fact that it has a perceptual 'pattern' to measure the randomness of a sequence of a binary string. In reviewing the literature on the notions of patterns in Kolmogorov complexity/Algorithmic Information Theory the real question arises, which patterns qualify for status as random, especially as a measure in a sequence of a binary string?

[1]   Knuth, D.E., The Art of Computer Programming: Volume 2 Semi numerical Algorithms (Addison-Wesley Publishers, Reading), 1997, p. 149.

[2]   Knuth, D.E., The Art of Computer Programming: Volume 2 Semi numerical Programming (Addison-Wesley Publishers, Reading), 1997, p. 169-170.

[3]   Shannon, C.E., Bell Labs. Tech. Jour. 27, (1948), 379-423 & 623- 656.

[4]   Li, M. and Vitanyi, P., An Introduction to Kolmogorov Complexity and Its Applications (Springer, New York), 1997, p. 186.

[5]   M. Ge, The New Encyclopedia Britannica (Encyclopedia Britannica, Chicago), 2005, p.637.

[6]   Martin-Lof, P., Infor. And Contr., 9.6 (1966), 602-619.

[7]   Uspensky, V.A., 'An introduction to the theory of kolmogorov complexity' edited by Watanabe, O. Kolmogorov Complexity and Computational Complexity (Springer-Verlag, Berlin), 1992,p. 87.

[8]   Tice, B.S., Formal Constraints to Formal Languages (AuthorHouse, Bloomington), in press.